The Poetic Species

A Conversation with
Edward O. Wilson and Robert Hass

Also by Edward O. Wilson

A Window on Eternity: A Biologist's Walk Through Gorongosa National Park

Letters to a Young Scientist

Why We Are Here: Mobile and the Spirit of a Southern City (with Alex Harris)

The Social Conquest of Earth

Kingdom of Ants: José Celestino Mutis and the Dawn of Natural History in the New World (with José M. Gómez Durán)

The Leafcutter Ants: Civilization by Instinct (with Bert Hölldobler)

Anthill: A Novel

The Superorganism: The Beauty, Elegance, and Strangeness of Insect Societies (with Bert Hölldobler)

The Creation: An Appeal to Save Life on Earth

Nature Revealed: Selected Writings, 1949–2006

From So Simple a Beginning: The Four Great Books of Charles Darwin

Pheidole *in the New World: A Dominant, Hyperdiverse Ant Genus*

The Future of Life

Biological Diversity: The Oldest Human Heritage

Consilience: The Unity of Knowledge

In Search of Nature

Naturalist

Journey to the Ants: A Story of Scientific Exploration
(with Bert Hölldobler)

The Diversity of Life

Success and Dominance in Ecosystems: The Case of the Social Insects

The Ants
(with Bert Hölldobler)

Biophilia

Promethean Fire: Reflections on the Origin of Mind
(with Charles J. Lumsden)

Genes, Mind, and Culture: The Coevolutionary Process
(with Charles J. Lumsden)

Caste and Ecology in the Social Insects
(with George F. Oster)

On Human Nature

Sociobiology: The New Synthesis

The Insect Societies

A Primer of Population Biology
(with William H. Bossert)

The Theory of Island Biogeography
(with Robert H. MacArthur)

Also by Robert Hass

Field Guide

Praise

Twentieth Century Pleasures: Prose on Poetry

Human Wishes

The Essential Haiku:
Versions of Basho, Buson, and Issa

Sun Under Wood

Now and Then: The Poet's Choice Columns 1997–2000

Time and Materials: Poems 1997–2005

The Apple Trees at Olema: New and Selected Poems

What Light Can Do:
Essays on Art, Imagination, and the Natural World

The Poetic Species

A Conversation with Edward O. Wilson and Robert Hass

EDWARD O. WILSON

AND

ROBERT HASS

Foreword by Lee Briccetti

BELLEVUE LITERARY PRESS

New York

First Published in the United States in 2014 by
Bellevue Literary Press, New York

For Information, Contact:
Bellevue Literary Press
NYU School of Medicine
550 First Avenue, OBV A612
New York, NY 10016

Printed on text paper certified by the Forest Stewardship Council.

Library of Congress Cataloging-in-Publication Data
is available from the publisher upon request.

Bellevue Literary Press would like to thank all its generous donors—individuals and foundations—for their support.

This publication is made possible by grants from:

The New York State Council on the Arts with the support of Governor Andrew Cuomo and the New York State Legislature

Book design and composition by Mulberry Tree Press, Inc.

Manufactured in the United States of America.

Printed on 30% post-consumer recycled paper.

FIRST EDITION

1 3 5 7 9 8 6 4 2

ISBN: 978-1-934137-72-7

The degree to which we are all involved in the control of the earth's life is just beginning to dawn on most of us, and it means another revolution for human thought.

—Lewis Thomas,
The Lives of a Cell, 1974

The Poetic Species

A Conversation with
Edward O. Wilson and Robert Hass

FOREWORD

O let them be left, wildness and wet;
Long live the weeds and the wilderness yet.

Gerard Manley Hopkins, "Inversnaid"

EDWARD O. WILSON has called *Homo sapiens* the poetic species because our cognitive infrastructure is dependent on analogy and associative thinking. As the longtime director of Poets House, a sixty-thousand-volume, open-access poetry library and a place for the poetic species to encounter—well, *poetry*—I was fascinated by Wilson's analysis. So Poets House queried the American Museum of Natural History about convening an evening to explore intersections between poetry and science, hoping to create a public dialogue between scientist Edward O. Wilson and former US poet laureate Robert Hass. The conversation, from which this book originally took shape, occurred on December 6, 2012, at the museum.

Wilson is one of the great field biologists of our time. Besides his international stature as an entomologist and as a thinker about the implications of evolution, his scores of books intended for a general audience have invited the public into a dialogue about the passion and creativity of scientific inquiry. Hass is a poet, scholar, and thinker of suppleness and reach, with a lifelong commitment to environmental issues. Both care deeply about the future of life on earth. One of the possibilities for exchange between these two brilliant writers had to do with the proximity between the language of science and literature. How are both rooted in observation and articulation? What is the promise of consilience?

The conversation took pleasure in various overlapping habitats of imagination and articulation; conservation and evolution. But ultimately it demonstrated a shared sense of urgency. Human behavior is changing the living world. We have come to a moment of environmental crisis with profound implications for our own species and for the planet we share with other forms of life. Extinction rates are exponentially accelerating, and it has been predicted that half the species on earth could be lost within the next century. This would be the worst degradation of biodiversity in millennia. The causes are a noxious mixture of habitat loss and economic exploitation, deforestation, and climate change.

Edward O. Wilson and Robert Hass are both deeply concerned about human impacts on biodiversity. Both seek ways to join emotion with rational analysis to create a deeper and more enduring conservation ethic (Wilson 1985, 119).

Prior to this conversation, Poets House had been engaged in a project entitled the Language of Conservation in six major city zoos, placing poems in prominent juxtaposition with the animals to test whether or not the poetic species might become engaged in deeper affiliations with other living creatures through poetry.

Millions of people throughout the country encountered the poems at the zoos—fragments; full texts; poems in translation from all over world, often from the place of origin of the animals. In exit interviews, we learned that visitors could remember many of the lines of poetry and that their conservation IQ was actually raised . . . but that they did not always know that what they liked was poetry.

This confirmed what Poets House had learned from years of work with public libraries and their communities: when people experience poetry, they are often surprised and delighted. But if you tell them that it is coming, they get nervous.

In the United States, this fear is something poetry shares with the sciences. Though practitioners are passionate, the general public initially responds with suspicion, needing to be invited in.

Though it surprised most curators and scientists participating in the zoo program, the poetry worked to engage people emotionally with what they were seeing. It worked not because visitors to the zoos were students of literature, but because poetic thinking and metaphor are intrinsic to the way the human mind operates and are critical to grasping abstraction (Deutscher, 117). Poetry, so rich in metaphor, activates imagination through a process that transfers affinities from dissimilar subjects to one another, creating new ways of seeing and feeling.

Human beings, of course, have a natural history. But psychologists, anthropologists, and cognitive scientists are just now beginning to map the brain and to describe human cognitive processes in relation to the hereditary nature of the human experience. New research suggests that one hundred thousand years ago, much earlier than previously thought, *Homo sapiens* had a spacious braincase with an estimated one hundred billion neurons processing information transmitted along 165,000 kilometers of myelinated nerve fibers across 0.15 quadrillion synapses (Pringle, 42). It is conjectured that a bigger brain enabled human beings to free associate and create analogies, while also making it possible for swift shifts to analytic thinking.

It was Wilson's 1975 book, *Sociobiology: The New Synthesis* that defined a threshold in the use of evolutionary theory to better understand human behavior. When it was published, he was treated as a pariah by colleagues at Harvard, called a fascist and a racist in the press, and even doused with water at a scientific conference. Happily, the days of the Inquisition and being burned at the stake are gone, or Wilson might not have been able to work so vigorously into his eighties.

Wilson's *The Social Conquest of Earth* created another ruckus in 2012, just before the conversation at the American Museum of Natural History took place. In that book, he broke ranks with those who hold to the theory of kin selection—to which he had subscribed over the vast majority of his career—and proposed group selection as an alternative. The group selection model based on mathematical computer analysis was developed between 2004 and 2010. Many scientists dispute the science. Many in the audience at the museum were uncertain about what they were hearing—was it being suggested that aggression between groups is biological predestination?

If group selection represents a scientific paradigm shift, it will take many years for the ambiguities of the theory to be worked out. Robert Hass, gracious and always curious, tried to parse the changes in Wilson's evolving thinking. But the discomfort in the audience of seven hundred living beings (not counting an extraordinary number of microbes and at least one fly on the stage) was, at times, palpable.

There was, nonetheless, much common ground. Hass is a cofounder of River of Words, an educational program that engages students in the study of watersheds through field observation, research, and the intellectual synthesis of creative writing. Both Hass and Wilson have also been involved with place-based learning that gets youngsters to identify place as a living entity. I know children in Vermont who were assigned a plot of ground that they could observe intimately over time, studying the interactions of flora and fauna. From the woods of Vermont to the watersheds of California to the abandoned parking lots of the inner city, this kind of study gets children to slow down and look at the world and to experience the poetry under their boots.

From the Paleolithic era, and probably before, the arts have been integral to imaginative mastery and the sense of connectedness to the living world. For our intensely social species, the arts can still be a powerful means of increasing empathy for one another and for the other species our cultures endanger. The imaginative ability to extend thinking to a wide range of possible life experience is adaptive (Dutton, 126).

At the December 2012 conversation, pleasure rippled through the audience as Robert Hass read from Elizabeth Bishop's "The Fish":

He was speckled with barnacles,
fine rosettes of lime,
and infested
with tiny white sea-lice,
and underneath two or three
rags of green weed hung down.
While his gills were breathing in
the terrible oxygen
—the frightening gills,
fresh and crisp with blood,
that can cut so badly—
I thought of the coarse white flesh
packed in like feathers,
the big bones and the little bones,
the dramatic reds and blacks
of his shiny entrails,
and the pink swim-bladder
like a big peony.

Both Hass and Wilson assert that the impasse between the humanities and sciences, dating since the Enlightenment, must be bridged if we are to work together to address the cataclysmal change in nature. Let us take pleasure in what we learn, aligning thinking with feeling.

Poetry is not purely associative. Science is not purely analytic. All skilled human production depends on subtle networks of cognitive capacities and the ability to transition between them. Let us use all the tools we have as we seek truth about our place in nature and our stewardship of life on the earth.

As electronic media changes access to information, museums, libraries, and cultural organizations are exploring new ways to work together to help their communities synthesize information. Perhaps the future of these institutions actually depends on finding common cause as facilitators of knowledge, redefining partnership as a value (Fraser, 123) as we consider the fate of the earth.

Much joint energy created this cooperative public space for generative thinking. Nancy Hechinger introduced me to Lisa Gugenheim, senior vice president of institutional advancement, strategic planning, and education at the American Museum of Natural History, and Bella Desai, director of public programs and exhibition education. I thank each of them, knowing that the museum's willingness to experiment was enabled—in the very best sense of the word—by the marvelous leadership of Ellen Futter, president of the American Museum of Natural History, and a dear Poets House friend, Laura Sillerman, who serves on the museum's board.

Finally, ever thanks to Robert Hass and Edward O. Wilson for their lifetime commitments to the news from the universe. For all of us, as the great Sufi poet, Rumi, has written: "Let the beauty we love be what we do."

—Lee Briccetti
Executive Director, Poets House

Bibliography

Deutscher, Guy. *The Unfolding of Language: An Evolutionary Tour of Mankind's Greatest Invention.* New York: Henry Holt, 2005.

Dutton, Denis. *The Art Instinct: Beauty, Pleasure, and Human Evolution.* New York: Bloomsbury Press, 2009.

Preston, Jane, ed. *The Language of Conservation: Poetry in Library and Zoo Collaborations.* New York: Poets House, 2013.

Pringle, Heather, "The Origins of Creativity," *Scientific American* 308 (2013): 36–43.

Wilson, Edward O. *Biophilia.* Cambridge, MA: Harvard University Press, 1986.

———. *The Social Conquest of Earth.* New York: W. W. Norton, 2012.

The Poetic Species

A Conversation with
Edward O. Wilson and Robert Hass

ROBERT HASS: I've been reading your recent book, *The Social Conquest of Earth*, and I knew that we were going to have this conversation, and the first question I thought to ask you was, how do you always manage to get in so much trouble?

Edward O. Wilson: Good scientists, like good innovators of any kind, are entrepreneurial, and they're the ones that are most likely to get into trouble.

And I've always enjoyed being in trouble. In science, trouble means progress. I got into a lot in the 1970s when I dared to propose in the book *Sociobiology* that there is such a thing as human nature. There was a time when social scientists as well as the American Left believed that there was no such thing as human nature and that there was no such thing as genetic influence on human behavior. Opposing these dogmas, I got into a lot of trouble. I once had to be escorted by police through the rear entrance of the Science Center at Harvard, because there was a mob out front calling for my head. Well, not exactly my head—just my discharge from Harvard. And other protesters in Harvard Square called for my dismissal for such a dangerous and outrageous idea. So yeah, I've been there. Now I'm bragging about how much I've been pilloried, and I bet you can't match that.

HASS: No, I can't. Poets tend not to get assaulted. They just get reviews they find grossly uncomprehending. And then they fume privately. But the question of human nature might be an interesting start to a conversation about science and the humanities. And especially about science and the humanities and the environment. It is still the case in the humanities—for the kids who are being educated now by the young generation of scholars—that there is a deep suspicion of the idea that there's any fundamental human nature. And it's worth saying that there are very good reasons, historically, for that suspicion; that is, in my territory, the poets and philosophers of the late eighteenth and early nineteenth centuries, of the Enlightenment and the romantic period, conceived of nature as a force in the constitution of the world of great simplicity and purity, a revolutionary force, and they set it against all the artifice and convention of a world run by a Louis XIV view of things. It was to be an enormously liberating and democratic idea. By the middle of the nineteenth century, people like Marx were saying that, while the idea of nature was once liberating, the bourgeoisie had taken it over to put women in their place, to justify slavery and imperialism, to say that homosexuality was perversion, to say that a monotheist male God was the natural order, et cetera, et cetera, so that the reasons for wariness about the idea of anything being absolutely determined by nature, culturally, have bled over into the ways we talk about it biologically.

WILSON: Yeah, it's always been a tough road for biologists. You're right, the Enlightenment died in the early 1800s. It died because the enthusiasts, many of whom were scientists, very early scientists, were not able to deliver. And, of course, the romantic movement took over. And then when something resembling the Enlightenment began to come back in the twentieth century, it came back in a rather—how shall I put it—in a friable and easily distorted form. It was that behaviors and abilities really are genetically determined, and from that came the birth of eugenics. And from eugenics came not just a fascist version, but also a Soviet version. For the first ten years of its existence, the USSR had institutions of eugenics designed to create the perfect socialist man. We've seen the disaster that eugenics brought. Little wonder the social scientists wanted to distance themselves. In the United States the countermovement owed a lot to Franz Boas. He was one of the great anthropologists of all time—but he was set pretty much in the view that culture is everything. And so the social sciences—under the Boasian worldview, stellar figures like Ruth Benedict and Margaret Mead—settled upon a no-instinct doctrine, and that's what I and a few others came up against in the 1970s.

HASS: When you immediately, upon the publication of *Sociobiology*, got accused of being a determinist, a eugenicist . . .

WILSON: Yeah, yeah, keep going!

HASS: I think I'll stop there.

On the literary and the philosophical side of things, this debate is about the question of free will, about the relation between human choice and the idea of fate. So many of the old stories are about fate being fulfilled or frustrated. It has always been an intense human fascination, how much freedom we have and whether we have any at all. I remember at a poetry reading in San Francisco once, during the question and answer period, an earnest young woman—she was quite pregnant,

I remember—raised her hand and asked if there was such a thing as free will. The old poet Kenneth Rexroth looked at her as if he were a little ashamed of himself for having given the impression that he could answer such a question, and then said, very kindly, "We can't know, and we have to act as if there is." I thought that was a good answer. And I was very struck by the way your answer to this question has evolved. Your book *Social Conquest* seems to want to be a sociobiology of human nature, or to use your term, a consilience on this issue. Am I right in thinking that the core of your argument about this, in this book, begins with the idea that the success of human evolution, the way our species has come to dominate the earth, had to do fundamentally with our being social, being members of a pack, and that there are two competing views of the evolution of social behaviors in human beings?

WILSON: You refer to two opposing views of how altruism and advanced social organization based on altruistic division of labor have evolved in animals. It's a rare event that has occurred, to the best of our knowledge, only nineteen times in the history of life, twenty if you include human beings. In every case, this level—seen in a very advanced form in ants and termites, for example—is approached along only one evolutionary path. It was necessary first for females of the species to build a nest that they protect from intruders and in which they raise the young to maturity by foraging and bringing food to them. It happened in humans at the *Homo habilis* level about three million years ago when these distant ancestors began to consume more meat through hunting and scavenging, leading to the invention of campsites, the equivalent of nests, and the altruistic division of labor. Meat was found in more than bite-size amounts, brought home and shared altruistically with those altruistically protecting the offspring of the hunters. One theory, to continue what is necessarily an oversimplification, holds that the cause of such advanced social evolution is kin selection—kin helping collateral kin, such as siblings and cousins, in addition to personal offspring. This sounds plausible, but, as I and a group of mathematicians and biologist experts on the subject have shown, it is not mathematically sound and doesn't accord well with the steps we know were involved in advanced social evolution. The new theory, multilevel natural selection, is sound at both levels, theory and fact. A key part of it is competition between groups. Competition *among* group members promotes rivalry and opposes altruistic cooperation; competition *between* groups promotes the opposite among members of each group. Group selection is an anathema to many kin selectionists, but it's easily demonstrated in nature.

HASS: One of the things that's terrifically interesting about this idea, that on the one hand . . . let me just trace that discussion a bit . . . In order to get the human brain large enough to be really good at throwing spears and eventually developing language, you have to get a very large head out of the birth canal—I say it that way while acknowledging that you and I haven't had to do it—and you have this prematurely-born mammal child that requires years of very close attention. And the other evolution that has to occur is the female has to shift out of an estrous cycle into a state of ongoing sexual interest and availability in order to secure a family unit that can protect this helpless being. So then, the core evolutionary object is, first of all, the family, and secondly, the larger social group, and the need to develop social behaviors in order to protect the ongoing life of the group is one powerful evolutionary impulse.

WILSON: I believe that's true, and what you've hit upon is the way the human brain evolved up to its present level across a scant three million years. It grew from four or five hundred cubic centimeters to fourteen hundred cubic centimeters in the present time. And in pushing to such a high level, it acquired a tremendous advantage over the rest of life on Earth. That achievement suggests that the meaning of humanity had to change, the point you've made. The life cycle and the whole manner of communication and organizing had to change. And out of that comes the necessity of group selection—group versus group. There had to be an instinctual emphasis not only on the family, and how well it functions as a group, but also on the society and how well it functions as a society. What you expect to see in history, and do find, is a lot of conflict, including the deadly violence that seems to be a hallmark of our species.

Hass: Yes, for poets it's always been interesting to notice that the culture that showed up when humans passed over the event horizon of writing was a male warrior culture. Though there is other stuff from the beginning—prayer, erotic desire. But back to the main idea. On the one hand, there is group selection, and on the other, in any individual member of the social group, there is also the ordinary impulse of not only mammals but all living organisms to selfishly produce and reproduce their own seed, no matter what.

So there's the selfish gene notion and there's the absolute necessity of group selection, and altruism?

WILSON: Correct, sir. And this is the reason, as I've argued, that we are constantly conflicted. It's our basic nature to be conflicted. There is no equilibrium between the consequences of group selection favoring altruism—group cooperative behavior—and selfish behavior within the group, trying to get the best we can within the group. And may I suggest that we've now come up to the borders of the humanities.

HASS: Well, we've come to a place where the dance of the tension between these two things must constitute something of what we mean by *consciousness*, by the experience of having choice and free will and moral life. A productive oscillation between a social self and a private, individual self, in each person and in the species.

WILSON: We scientists can gabble on forever about this subject and gather data sets and maybe devise new theories. Eventually we will have a pretty well-knit body of scientific thinking about this subject. Now, what will that mean to the humanities, if anything? It's a tough question, but you just told me about my field, so probably I'm supposed to tell you about your field.

HASS: I would be grateful if you would. And you have, in a way, by writing about the evolution of culture. One of the interesting things about this idea is that it has so many echoes in art making. Artists almost always start with a kind of play based on elements that are fixed and variable, things that conventions express, set forms in music, set patterns in comedy, fixed rhythms in poetry, on the one hand, and, on the other, departures from those conventions that lead to new ways of seeing and feeling. In a way, it's the same oscillation, between sensations that make us feel safe, part of the group, and sensations that make us feel free and on our own. The formal imagination in art—the half-conscious shaping that occurs when an artist is at work—is always working on this problem.

Working also at the level of content, at how, in large ways, we imagine the world.

This is probably the point at which this conversation bears on the environment and where the human species is now. I teach with a friend who's a geochemist and a hydrologist an introductory environmental studies class for large groups of Berkeley freshmen, and the first two pieces of reading we give them are John Muir's account (which Teddy Roosevelt read) of a really ferocious Sierra storm. Muir climbed the highest lodgepole pine he could find and strapped himself to the trunk so that he could enjoy the storm and sway back and forth, conducting it like a Beethoven symphony. He uses the occasion of the essay to say that it is mistaken to think that Darwin's proposal about the "survival of the fittest" is cruel. Rather, he says, the storm is strengthening the forest, that he is witnessing the love lavished on the planet by forces of nature that dazzle us, improve the resilience of the forest, and make a grand symphonic music out of the carbon cycle. So there's this ecstatic, delicious essay.

And then we give them an essay by Joan Didion about how people in LA feel when the Santa Ana winds blow through. One of the first things that happens is the homicide rate goes up, especially among married couples. I think she says it doubles every time those dry, maddening winds riffle through the pretty neighborhoods. So you get on the one hand this LA noir view of us as totally determined beings, just Pavlovian and rather sinister bundles of responses to the weather, and on the other this incredible celebration of nature.

So here's another question. Do these two forms of imagination have a correspondence to the ways imagination works in science? I've read that one of the pioneering American ecologists, Frederic Clements, proposed that, after Darwin, the way to study nature was to study the rich interactions among species in a given place, that it made sense to talk about a pond or a dune system as a community, even as a superorganism. This was, I guess, in the 1910s, during the Progressive Era. Clements was a Nebraskan who went to school with Willa Cather, so he proposed a sort of midwestern Socialist nature. And then a decade or so later, another biologist, Henry Gleason, published a paper that said Clements's ideas were nonsense. I believe the essay is called "The Individualistic Concept of the Plant Association." And it argued that nature is ruthless competition among species. A 1920s stock market nature, hungry, unstable, and directionless, but linear and irreversible in its outcomes, and that the idea that there's any kind of cooperative interaction is a false paradigm. So it would seem that this argument between the selfish gene and the altruistic gene has gone on at every level of thinking about nature.

WILSON: Yes, genes are usually conceived as selfish in biology when we focus on natural selection arising from competition among members of the same group. Altruistic genes are conceived as those that oppose selfish genes by promoting cooperation that requires personal sacrifice among members of the same group. Yet, in one sense both kinds of gene are selfish. They differ only by the level of biological organization on which in each case natural selection acts. That is, individual "selfish" genes spread by competition of individual versus individual in the group, and "altruistic" genes in the individuals arise by competition among the different groups to which the individuals respectively belong.

Let me shift from this topic to the nature of the humanities viewed from the perspective of biology. I've suggested many times that the humanities, and especially the creative arts, are the natural history of *Homo sapiens*. The descriptions based on them describe the human condition and human nature in exquisite detail, over and over again in countless situations. When verbal descriptions are novel in style and obedient to the most basic principles of human nature, when they connect old memories, create new images, and stir emotions all together, we call that great literature. The important innovator produces a tableau of relationships in a story that describes not just the particularities of a place in time, but something that is true for humanity as a whole for all time. If the humanities are tightly prescribed in literature and the other creative arts by the unique properties of the one species, *Homo sapiens*, as appears to be the case, will they ever go beyond that? The same can be asked of the humanities as a whole. Are they in danger of becoming biological, or what? What's going on?

Hass: You've written about the evolutionary origin of culture, and on the evolutionary origin of the arts, and what their adaptive value is, which would tell us something about what we do when we listen to stories or listen to music, or might tell me something about what I do when I try to make poems. Because neither the social imperative nor the selfish gene by themselves or together quite account for the existence of imagination. You could say language evolved out of the social imperative. And then the selfish gene led individuals to want to excel in the powerful uses of language and art making and music making and dance. So the social drive shaped the uses of imagination. It made it possible for humans to share their invisible inner worlds with each other. I often think of this in relation to dreams. Once they could speak, humans could tell each other their dreams. They could find out that everybody has dreams, that there is this parallel world of meaning-making or traveling that goes on in the resting mind. This would have led to shamanism and to a whole magical and metaphorical vocabulary.

WILSON: Good point. We dream together, and as a result the cultural products of human nature are vastly expanded and enriched. And approaching from the other side of the divide, biology progresses and connects with the humanities. What biology seems to be doing at the moment is to reveal the roots of ambiguity that define human nature. We've been talking, for example, about the eternal confliction of the human mind, between self-serving behavior for the individual and for its offspring, versus service to the group. This clash of evolutionary forces can never result in an equilibrium. If it goes too far toward individualism, societies would dissolve. If, on the other hand, it goes too far toward obedience to the group, the group would turn into an ant colony. So, we're creatively conflicted, moving back and forth between sin and virtue, rebellion and loyalty, love and hate.

The pull evolves by individual-level natural selection operating on members of the group. It draws the mind toward self-centered, even selfish behavior, which in strong forms we call sin. The opposing pull by group-level natural selection draws the mind in the opposite direction toward group-centered behavior, which we call virtue. The inner mind struggles constantly with this hereditary dynamism. It's the story of our personal lives. Which we share, with one another, as you say. The creative arts are the sharing of our inner desires and humanity's struggle. The humanities are our way of understanding and managing the conflict between the two levels that created *Homo sapiens*. The conflict can never be resolved. And we shouldn't try too hard to reach a resolution. It defines our species and is the fountain of our creativity.

HASS: American poets have struggled with this. Robert Frost has a very dire view of nature, of the darkness of it, though he posed as a genial poet. And Wallace Stevens thought that it was basically unknowable. He wrote a lot about . . . well, no, he wrote a lot and playfully from the condition of not knowing whether our sense of order and beauty came from the world or from the mind. In my reading he seemed to think that humans learn everything they learn about order and beauty from the natural world without ever really knowing what it is. I remember the first time you and I had a conversation in public, you said, "Every species lives in its own sensory world."

Wilson: Well, yes. It's easy to forget that after all is said and done, humanity is a biological species in a biological world. Every species is different, each is exquisitely well adapted in one way or another to the biosphere.

Hass: I was very struck by it. Stevens in one of his last poems says he imagines as a kind of final act of nature a bird singing "without human meaning, without human feeling, a foreign song." The idea that every creature has its own reality scared poets at the beginning of the twentieth century, made some of them feel we were groping blindly—it in effect kicked us out of a comfortable anthropocentric community—but it also allowed some modern poets this sense of absolute mystery at the core of existence. It came of knowing that we would never know exactly what a bird's experience is, or what an ant's experience is. It has been an unhousing of the imagination, and it was brought on by the thrust of science to be at home in the world by understanding it. It said we move among great powers and mysteries and only glimpse their meanings, the meaning of what it's like to be another creature, and therefore also the meaning of being a self, a person. It's as if, once again, we were moving among great powers and mysteries—something like the view of the world we glimpse in early human culture—a kind of polytheism of amazing powers. The first time I ever went into one of the tombs in Cairo and saw these three-thousand-year-old carvings of birds and fish, absolutely accurately rendered, the smallest little hump on the crow's back, which is unlike any American crow, all of it rendered in absolute detail—I thought, humans have been paying attention to these things and fiddling with these powers from the beginning. Science, partly by the kind of patient observation that noticed the hump on the Nile crow's back and partly by leaps of imagination and by shared testing and dialogue, has made enormous progress in understanding certain things about the world, but the skill of those artists made me feel that we have always been pretty much in the same place with the same kind of knowledge and the same pull back and forth between ways of seeing.

WILSON: One thing biology has brought us is an idea of what the Germans call the *Umwelt*, the sensory world around us, and how peculiar the human mind is. We think we see everything, but that's a delusion. We see, for example, only a tiny sliver of the electromagnetic spectrum. And we think we see four colors, but that, you know, is the way the optical lobe is organized, and the sensory receptors react to frequencies. Frequency impulses of the spectrum are coded, coming back from the retina. Other animals see infrared that we can't see, and others see ultraviolet that we can't see, and then even more, we only hear a tiny sliver of the frequencies in other modalities that animals are using all the time. If you go out into the rain forest, as I did recently in Africa, and you turn high frequency receptors, as a colleague did with a special apparatus, you can hear dozens of species of katydids singing that you didn't know were there. Almost all creatures, incidentally, live by smell and taste. Those are what they sense superbly well. It's an entirely different universe from the one we live in. Humans, in contrast, are among the very few audiovisual creatures on the planet. Birds are another example, and perhaps that is why we love them so. I don't know what you make of that! Maybe the knowledge interpreted together by science and the humanities will tell us what the real life of an ant would be, or a katydid, or a wolf. It's an interesting thought.

But now, if I may move on to something you brought up that is so important, where I think you and I are on the same side: a disturbing trend, the habit of diluting or even dismissing nature in our thinking. In what may be called the anthropocene philosophy, some writers suggest we ought to give up on the idea of wilderness, that it's just in our head, and somehow we should stop worrying about preserves and begin thinking more about mingling nature and humanity together into some sort of Edenic garden and save biodiversity that way. I am frightened by this. And it's gotten support by a few people calling themselves conservationists. Is there a serious risk here?

Hass: Oh, yes. I know you know quite well Aldo Leopold's take on this, and you've been one of his heirs. What he argued in 1949 was that human ethics—in your language it would certainly be one of the main developments in the social imperative, in rules for community bonding—needed to extend its range of concern to the rest of life. And you seemed to be proposing an evolutionary ground for that in the idea of biophilia, of—would you call it an instinct?—anyway, a disposition of animals to be fascinated by animals. I've been struck by the way that you've thought about this in relation to, or at the same time that you were thinking about, war and human violence, the long thought that you have had about the relationship between the evolutionary power of the group and tribalism and religion and war; that the really dangerous and dark side of our social nature is that it is always creating an "us against a bad them" that links to the incessant and endless violence in human history. So, I don't know what you can say to that, in relation to what you've called a New Enlightenment. That seems to imply a hope that, in relation to the kinds of mystification of power that are involved in tribalism and in certain kinds of organized religion, we need to do the work of disenchantment that science does. And the interesting thing about this is there is a familiar discourse in environmentalist humanities, which is that science, by objectifying the natural world, has killed spirit in it and allowed industrial civilization to manipulate it at their will.

Wilson: Well, there's a lot to that as well, for sure.

Hass: There's a younger group of environmental and cultural historians who have been arguing, on the contrary, that the Old World's habit of seeing animals in ways that immediately turned them into metaphors for human things—lions are a pride, snakes are an evil subtlety—I'm thinking of my colleague Joanna Picciotto, who has argued that in the seventeenth century with the development of empirical science, people started saying: "How does it feel to be an ant?" "What would it be like to be a spider?" That on the contrary, the sciences began to look with an earnest eye at how creatures other than humans made their living and didn't see them as symbols of human attitudes and conditions. That the natural sciences began to extend the community of concern to the rest of life by trying to see animals and plants on their own terms.

This connects to our other conversation. Darwin and Mendel, the whole study of organisms in their environments and the development of genetics has given us the richness of the gene pool we came from and what's happening to it to contemplate. In both kinds of work, literary and scientific, we have been struggling with it—postmodern poets trying to figure out how to write about how little there is left and what in the literary and philosophical traditions have contributed to the devastation. That's another form skepticism in the humanities takes. And scientists are having meetings with titles like "Which species can we still save?" In my part of the world, what I am aware of is the very different ways in which people have tried to assimilate the idea that "nature is over." Not to attack Bill McKibben's wonderful book, but the thesis of that book is that, to the extent that we mean by nature, natural processes in the world untouched by human beings, then—between acid rain and global warming—there is nothing left untouched by human beings. "Nature is over"; we have become what this said we were, in charge of the garden of the world. And that's true enough.

WILSON: That is chillingly accurate in its description of what I classify as defeatism. And I am hearing this from people who are, you know, pretty well educated.

HASS: And the next part of the argument that I hear is that because it's the garden, we might as well give up on the idea of wilderness preservation, because wilderness is, after all, a cultural concept. It's selling rain-forest ice cream to people in cities as propaganda to take land away from peasants who want to cut down trees and start raising cattle and making money in the rain forest.

WILSON: That gives me an opportunity to confess: I am an extremist. I believe in wildernesses. I've been there. I've studied thousands of species living there, in ecosystems much the same as they were millions of years ago. I believe, I think, in reference to the species that we might still save—and a growing number of them are endangered—that we need parks, big ones, lots more of them. I think we should be thinking about giving a large part of the world's surface to wild land. To do so is not just being a conservationist—not just saving species—we must hold on to the rest of life. I sometimes hear, "Well, you know, we may need to have a triage. There are certain species that are just so endangered and natural areas that are being encroached on in the interest of the economy, and the needs of the people around them, we'll have to let them go." Well, I have a doctrine: Save Them All. I don't mean to make a political statement. I'm making a moral statement. We have to develop a new and better ethic to save the rest of life.

I'm spending more and more of the time I have left on national parks. I visit and consult on Mozambique's Gorongosa National Park, which was badly damaged during the 1977–92 civil war. I've recently traveled to the South Pacific to study national parks and reserves there. In the course of this work, I've been supporting and collaborating in a complete survey of the animal species, including a vast array of insects and other arthropods. I'm also part of a group from around my hometown of Mobile, Alabama, working to create a new national park there. It will be very large for the eastern United States, joining the wilderness of the Mobile-Tensaw River Delta, the second largest delta in the United States, plus the deeply dissected ridges and ravines of the Red Hills directly to the north. Polls suggest a large majority of people in the Mobile area support it, and we have friendly signals from the National Park Service. If the Mobile park becomes a reality, it will protect the largest number of plant and animal species of all the American national parks. This whole initiative, here and abroad, turns out to have serious political implications. We need a strong ethic and a renewal of the Rooseveltian spirit of creating and holding on to national parks.

HASS: I also just think it's a tactical matter that people who care about these things should not always be in the position of defending against the next development; it is just a much more powerful thing to be proactive in creating new places . . .

WILSON: You're quite right about that.

HASS: A metaphor for this, and also a fact, which is lovely and also terrible in this case, is the demilitarized zone between North and South Korea, which I got to visit recently. It's about three miles wide and runs from one end of the peninsula to the other, and it's the most demilitarized piece of real estate on the earth. There are, as you know, eight species of Asian cranes. All of them are either endangered or nearly endangered, and two of them are making a comeback because they do their winter foraging in the demilitarized zone, which has been made into an unintentional national park. And if North and South Korea ever settle their problems and remove the last nuclear trip wire of the Cold War, that land will probably be developed, and those two species, which are ten million years old, will be gone from the earth.

Wilson: Wouldn't that be a wonderful development if it were made into a park? I've been associated with South Koreans and Korean Americans since the beginning of the DMZ peace park initiative. Some North Koreans actually have very quietly expressed interest in it. It would have a lot of economic value. And you are quite right; and did you know the leopards have returned?

HASS: And the other part is thinking about the experience of the storm, Sandy. I think about the Hudson River and the East River. And your part of the Atlantic Ocean, and how following the work you do, we educate our kids so they become stewards of those places. That they have the poetry and they have the science in their heads to become stewards of biologically alive rivers and of our remnant wild places. Since it looks less and less likely that this country is going to address our use of fossil fuels, it's almost a certainty that the next couple of generations are going to be coping with the consequences. If we can't change, maybe we can educate the next generation, which will be forced to.

Wilson: You know, you just said something that's profoundly important. And it is that in terms of conservation . . . we've been on the defensive too long. Countless times I've been asked in serious surroundings, "Well, what good is biodiversity? Why should we keep all these bugs and snakes and God knows what else?" And then I find myself, like a witness in court testifying for some wretched felon, reciting all the reasons why I think biodiversity is important, and it's just dispiriting. I think if we could now take the offensive, in many ways, you know, victory may be in sight, because this is important. This is saving the rest of life, this is saving the habitat in the world in which our species was born. And I think we ought to feel pretty positive about it and pound the lectern more. Teddy Roosevelt, where are you?

HASS: One kind of work like that really is education, and in the way that you describe our social evolution, and we talk about the evolution of language, and the evolution of music, and the evolution of dance being part of this altruist and communitarian side of our genetic inheritance—that people learn to sway together, sing together, tell certain types of stories that they share . . . if our kids are not being told those stories, if the children of New York are not wandering through here and seeing those North American mammals and feeling some kind of wonder in their presence, if they don't know the names of the trees in Central Park, then how are they going to be stewards or caretakers of them?

Wilson: Do you foresee in literature, in letters, fiction, especially, the possibility of an infusion of more biological knowledge having to do with the properties of the rest of life and nature—real nature? Do you see any trend or prospect of that? Or at least more than exists today?

Hass: I think we have to work at it. Wonder is one place to start. I was asked to go to my granddaughter's kindergarten class and to talk about poetry. And I didn't know if I would know how to do it, but I brought the book I had with me—which was the collected *Poems* by Elizabeth Bishop, and there is a poem of hers called "The Fish," and it begins, "I caught a tremendous fish." So I opened the book and said to these little kids, "Just say this poem with me, okay? 'I caught a tremendous fish,'" and this group of kids all on the floor looked up at me and said, "I caught a tremendous fish." And—I simplified the imagery a bit—I said, "It was very old and its skin," and they said, "It was very old and its skin," and I said, "Looked like roses on old wallpaper." And they said, "Ooh."

And I thought, this is a cinch. One kid raised his hand, and I said, "Yes, what's your name?" He said, "Rodney." And I said, "Yes, Rodney?" And he said, "I have to go to the bathroom." And I thought, education is not so simple.

Wilson: Oh, but what you were just doing was, you were telling a story in the best old style.

Hass: So, what stories do we tell? And what do stories tell?

Wilson: That we'll have to work on together. Education shouldn't be a matter of saying, "Okay, kids, listen up! Here's what you've got to know about the elementary biology and so on and so on, here are the basic facts, and the terms you need to know." It's somehow got to be made into a story. I don't know exactly how to do that, but maybe that's where the humanities can have a powerful new role.

HASS: Or just someone's imagination. Or many different imaginations, as we work out what to do about the world we are changing so radically, the world that will kill us and that gives us life. What to do to extend the social imperative you speak of to the rest of life? Do you have a recommendation for poets?

WILSON: Colonize science.

HASS: In his essay collection *Outside Stories*, Eliot Weinberger writes: "Art, said Louis Sullivan, does not fulfill desire, it creates desire." This is a very mysterious subject. I think both things are true. There are desires that art satisfies. I think that's what William Carlos Williams must have had in mind when he said that people are dying every day for lack of what can be found in poetry. At a very simple level, poetry can give us someone to talk to. To read is, in that way, to have your inner life acknowledged by somebody else's. As for the way in which art creates desire, I guess that's everywhere. Is there anyone who hasn't come out of a movie or a play or a concert filled with an unnameable hunger? It's striking that Eliot Weinberger is quoting Louis Sullivan. To stand in front of one of his buildings and look up, or in front, say, of the facade of Notre Dame, is both to have a hunger satisfied that you maybe didn't know you had, and also to have a new hunger awakened in you. I say "unnameable," but there's a certain kind of balance achieved in certain works of art that feels like satiety, a place to rest, and there are others that are like a tear in the cosmos, that open up something raw in us, wonder or terror or longing. I suppose that's why people who write about aesthetics want to distinguish between the beautiful and sublime. It's interesting in this way to think about the pleasure of a scientific or mathematical solution. On the one hand, it supposedly lays a question to rest. On the other, it can be beautiful—"elegant" is what people often say about an explanation that's also a solution. Beauty sends out ripples, like a pebble tossed in a pond, and the ripples as they spread seem to evoke among other things a stirring of curiosity. The aesthetic effect of a Vermeer painting is a bit like that. Some paradox of stillness and motion. Desire appeased and awakened.

Wilson: You speak for science as well. I've often referred to science and art having the same creative wellspring, which I believe can be expressed aphoristically: the ideal scientist thinks like a poet and works like a bookkeeper.

Hass: Ezra Pound described poetry as primary research in language. That's one way in which poetry and science are analogous practices. Though this would be especially true of poetry when you didn't have at the outset the faintest idea of where you were going. And probably a scientist, beginning on an experimental procedure, does know where she's going procedurally. She just doesn't know what the results are. Something akin there. And the other thing, of course, is observation leading to questions leading to methods of answering the questions. And another way, I suppose, would be that you can track that process quite diligently and still find that the critical insight—intuition—answer—comes to you out of nowhere.

APPENDIX

A short list of great classic literary works of science or nature writing and what these books in particular reveal about our evolving attitudes toward science

Robert Hass:

You know, I'd start near the beginning. Ptolemy's *Almagest*, Euclid's *Elements.* I had an odd education that involved a certain amount of the history of mathematics, so I would add the *Conics* of Apollonius to the list and Descartes's *Geometry.* For literary wonders, there's Pliny's *Natural History,* Lucretius's amazing poem, *De Rerum Natura.* A shorthand approach is a wonderful book edited some years ago by a classical scholar, Robert Torrance, entitled *Encompassing Nature*. It hits a lot of the high notes and is a feast.

And then the letters of Galileo and *Origin of Species.*

And then the American nature writing tradition. I suppose everyone would start with William Bartram's *Travels* and with *Walden.* And later on in the list (ignoring the prescription) would come, among others, Edward Wilson's *Naturalist* (especially interesting—to an outsider—for what a career in science is like) and his *Biophilia*.

And randomly, from the English natural history tradition, I love J. A. Baker's *The Peregrine* and, from Europe and the nineteenth century, Jean-Henri Farbre on spiders. Which makes me think to add Czeslaw Milosz's great poem, "From the Rising of the Sun," for the section "Diary of a Naturalist," which may be the twentieth century's most definitive expression of revulsion at the cruelty of the food chain.

Edward O. Wilson:

It's not a good idea to ask a scientist what he has read as literature, and it's a rare scientist who has wide experience in this creative art. (I'm not one of them who also reads widely in literature for its own sake.) But here are a couple anyway:

- Darwin's four great books (*Voyage of the Beagle, The Origin of Species, The Descent of Man*, and *The Expression of Emotions in Man and Animals*), each classic, seminal in its field, and beautifully written.

- Erwin Schrödinger's *What Is Life?*, proposing a connection between physics and genetics, stirred me as a teenager before the dawn of molecular biology.

- *Kingdom of Ants: José Celestino Mutis and The Dawn of Natural History in the New World*, which I composed with the Spanish scholar José M. Gómez Durán. I mention it here not because we wrote it or it has special literary merit, but because Mutis, the great priest and polymath (1732–1808), initiated scientific natural history in the New World independently of William Bartram. The book he wrote, which would secure his place in history, was lost. Over two centuries later, Gómez Durán and I wrote his book again, using his notes and correspondence newly discovered in Spain. This resurrection is the first account of Mutis's research in either Spanish or English, a major addition to the history of scientific reporting, and I think it worthy to cite here.

THE FISH

Elizabeth Bishop

I caught a tremendous fish
and held him beside the boat
half out of water, with my hook
fast in a corner of his mouth.
He didn't fight.
He hadn't fought at all.
He hung a grunting weight,
battered and venerable
and homely. Here and there
his brown skin hung in strips

like ancient wallpaper,
and its pattern of darker brown
was like wallpaper:
shapes like full-blown roses
stained and lost through age.
He was speckled with barnacles,
fine rosettes of lime,
and infested
with tiny white sea-lice,
and underneath two or three
rags of green weed hung down.
While his gills were breathing in
the terrible oxygen
—the frightening gills,
fresh and crisp with blood,
that can cut so badly—
I thought of the coarse white flesh
packed in like feathers,

the big bones and the little bones,
the dramatic reds and blacks
of his shiny entrails,
and the pink swim-bladder
like a big peony.
I looked into his eyes
which were far larger than mine
but shallower, and yellowed,
the irises backed and packed
with tarnished tinfoil
seen through the lenses
of old scratched isinglass.
They shifted a little, but not
to return my stare.
—It was more like the tipping
of an object toward the light.
I admired his sullen face,
the mechanism of his jaw,

and then I saw
that from his lower lip
—if you could call it a lip—
grim, wet, and weaponlike,
hung five old pieces of fish-line,
or four and a wire leader
with the swivel still attached,
with all their five big hooks
grown firmly in his mouth.
A green line, frayed at the end
where he broke it, two heavier lines,
and a fine black thread
still crimped from the strain and snap
when it broke and he got away.
Like medals with their ribbons
frayed and wavering,
a five-haired beard of wisdom
trailing from his aching jaw.

I stared and stared
and victory filled up
the little rented boat,
from the pool of bilge
where oil had spread a rainbow
around the rusted engine
to the bailer rusted orange,
the sun-cracked thwarts,
the oarlocks on their strings,
the gunnels—until everything
was rainbow, rainbow, rainbow!
And I let the fish go.

Acknowledgments

Bellevue Literary Press gratefully acknowledges Poets House and the American Museum of Natural History for co-sponsoring the conversation between Robert Hass and Edward O. Wilson on December 6, 2012.

Poets House is a sixty-thousand-volume poetry library and a national literary center that invites poets and the public to step into the living tradition of poetry. Poets House—through its poetry resources, literary events for adults and children, and exhibitions—documents the wealth and diversity of modern poetry and stimulates public dialogue on issues related to poetry in culture. Founded in 1985 by two-time Poet Laureate Stanley Kunitz and legendary arts administrator Elizabeth Kray, the library creates a home for all who read and write poetry. In 2009, Poets House moved into its permanent home, at 10 River Terrace in Battery Park City, on the banks of the Hudson River. Last year, sixty-five thousand people visited Poets House, and over five million were introduced to Poets House via its new website and national partnerships. For more information about events, and Poets House in general, visit poetshouse.org and join Poets House on Facebook and Twitter.

The American Museum of Natural History, founded in 1869, is one of the world's preeminent scientific, educational, and cultural institutions. The Museum encompasses forty-five permanent exhibition halls, including the Rose Center for Earth and Space and the Hayden Planetarium, as well as galleries for temporary exhibitions. It is home to the Theodore Roosevelt Memorial, New York State's official memorial to its thirty-third governor and the nation's twenty-sixth president, and a tribute to Roosevelt's enduring legacy of conservation. The Museum's five active research divisions and three cross-disciplinary centers support two hundred scientists, whose work draws on a world-class permanent collection of more than thirty-two million specimens and artifacts, as well as specialized collections for frozen tissue and genomic and astrophysical data, and one of the largest natural history libraries in the world. Through its Richard Gilder Graduate School, it is the only American museum authorized to grant the PhD degree. In 2012, the Museum began offering a pilot Master of Arts in Teaching program with a specialization in Earth Science. Approximately five million visitors from around the world came to the Museum last year, and its exhibitions and Space Shows can be seen in venues on five continents. The Museum's website and collection of apps for mobile devices extend its collections, exhibitions, and educational programs to millions more beyond its walls. Visit amnh.org for more information, become a fan of the Museum on Facebook, or follow us on Twitter.

About the Authors

Edward O. Wilson is an American biologist, researcher, theorist, naturalist, and author who is often dubbed "the father of sociobiology." His biological specialty is myrmecology, the study of ants, on which he is considered to be the world's leading authority. Wilson is the author of the *New York Times* bestsellers *The Social Conquest of Earth and Anthill: A Novel,* as well as the Pulitzer Prize–winning *On Human Nature* and (with Bert Hölldobler) *The Ants.* For his contributions in science and conservation, he has received more than one hundred awards from around the world. A professor emeritus at Harvard University, he lives in Lexington, Massachusetts.

Robert Hass, whose work is rooted in the landscapes of his native Northern California, has served as United States Poet Laureate and as a Chancellor of the Academy of American Poets. For Hass, everything is connected. When he works to heighten literacy, he is also working to promote awareness about the environment. He has published many books of poetry including *Field Guide, Praise, Human Wishes, Sun Under Wood, The Apple Trees at Olema: New and Selected Poems,* and *Time and Materials*, winner of both the Pulitzer Prize and the National Book Award. He is also the author of the essay collections *Twentieth Century Pleasures and What Light Can Do: Essays on Art, Imagination, and the Natural World*, winner of the PEN/Diamonstein-Spielvogel Award. Awarded the MacArthur "Genius" Fellowship, the National Book Critics Circle Award (twice), and the Yale Younger Poets Prize, Hass is co-founder of River of Words, an environmental education program for children, and a professor of English at the University of California at Berkeley.

Bellevue Literary Press has been publishing prize-winning books since 2007 and is the first and only nonprofit press dedicated to literary fiction and nonfiction at the intersection of the arts and sciences. We believe that science and literature are natural companions for understanding the human experience. Our ultimate goal is to promote science literacy in unaccustomed ways and offer new tools for thinking about our world.

To support our press and its mission, and for our full catalogue of published titles, please visit us at blpress.org.

Bellevue Literary Press
New York